RAIN FOREST AT NIGHT

RAIN FORESTS

Lynn Stone

The Rourke Corporation, Inc.
Vero Beach, Florida 32964

Printed in the U.S.A.

PHOTO CREDITS
All photos © Lynn M. Stone except p. 12 and 18 © James H. Carmichael

Library of Congress Cataloging-in-Publication Data

Stone, Lynn M.
   The rain forest at night / by Lynn M. Stone
     p. cm. — (Discovering the rain forest)
   Includes index
   ISBN 0-86593-396-0
   1. Rain forest fauna—Juvenile literature. 2. Nocturnal animals—Tropics—Juvenile literature. [1. Rain forest animals. 2. Nocturnal animals.]
I. Title   II. series: Stone, Lynn M. Discovering the rain forest
QL112.S723 1994
591.909'52—dc20   94-20907
                                 CIP

Printed in the USA   AC

TABLE OF CONTENTS

THE RAIN FOREST AT NIGHT

Little moonlight creeps past the web of leaves and branches that make the **canopy**, or roof, of the tropical rain forest. Night is inky black, like the inside of a deep, dark cave.

But the forest doesn't sleep at night. When the daytime animals hide, the **nocturnal**, or nighttime animals appear.

Nocturnal animals have special ways to survive in this wet world of danger and darkness.

The smoky jungle frog leaves its burrow at night to catch insects

HOOTS AND HOWLS

Rain forests are quiet places. Even the rain itself only whispers as it trickles through branches to the forest floor.

At night the silence is broken by animals calling or moving about in the branches or litter of leaves.

Evening rains stir frogs and toads into calling for mates. They chirp, croak, bark and trill. Mosquitoes buzz at night and cicadas hum. Monkeys howl and owls hoot.

Warm, wet nights in the tropical rain forest are filled with the calls of frogs

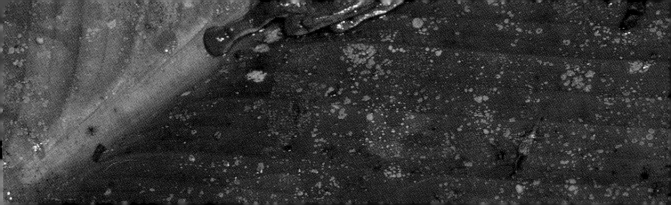

MOTHS

Most **species**, or kinds, of moths fly at night. Moths are closely related to butterflies, which fly by day.

Like owls, moths fly on soft, muffled wings. Quiet flight helps moths hear. Some species have such keen hearing they can escape from bats. These moths can hear the squeaks that bats send out to locate their **prey,** or food, which includes moths.

Some moths also have a keen sense of smell. A distant male moth can find a female from her scent.

Moths fly at night on silent wings

LEAF-CUTTING ANTS

Moths are joined by other nocturnal insects in the damp rain forests. Mantises, "walkingsticks," weevils, roaches, fireflies, katydids, crickets and leaf-cutting ants are all busy at night.

The amazing leaf-cutting ants haul pieces of leaves over tiny trails to their burrows.The ants chew the leaves, then spit them out. They add a body liquid to the leaf matter.

Fungi, a plantlike growth, soon appears on the leaf material. The fungi, not the leaves, becomes the ants' food.

Lit by the camera flash, leaf-cutting ants haul a leaf down a rain forest tree trunk

The kinkajou hides by day and prowls the rain forest canopy at night

The katydid's green cloak helps it hide during the day

SPIDERS

Spiders and other eight-legged animals, the **arachnids**, are active in the rain forests at night. Scorpions, tarantulas and wolf spiders are some of the eight-legged creatures that hunt at night.

The "furry" wolf spider has as many eyes as legs! Having eight eyes helps the spider find prey, mostly insects, in poor light.

Busy in the darkness of a rain forest at night, a spider prepares to eat a mantis

FROGS AND TOADS

Most species of toads and tree frogs are nocturnal. Females locate males by hearing their calls.

Under the cover of darkness, frogs and toads lay their eggs in rain forest pools.

Some species of tree frogs have large eyes. Their night vision is probably quite good, but their sense of touch is better. Tree frogs catch some insects when the insects step on them!

16   *The large eyes of the red-eyed tree frog help it catch insects at night*

SNAKES

Darkness hides hunters, and it hides the hunted. For certain rain forest snakes, like the "pit vipers," being a night hunter is no problem.

The pit vipers are a large group of poisonous snakes, including the fer de lance and several other species.

A pit viper doesn't need to see or hear its prey. The snake has an organ that senses heat. It can locate a mouse from the mouse's body heat.

A deadly eyelash viper patiently waits for prey in the stillness of night

BATS

Hundreds of species of bats live in rain forests. By day they hang like furry pears in hollow trees and caves or under broad leaves.

Most bats fly at night. They catch insects by sending out high-pitched squeaks. When the bat's squeak strikes an object, it echoes, or bounces, back to the bat.

By "reading" the echo, a bat finds its way—and its prey.

Their hunt over, these rain forest bats have made a "tent" for themselves with leaves.

BIG CATS

Many furry, four-footed rain forest animals are active at night. Sloths munch leaves and the big-eyed **lorises** hunt insects and fruit in the canopy.

Below the forest canopy, big, wild cats walk silently as they listen and sniff for prey. The biggest of the rain forest cats are the jaguars of South America and the extremely rare tigers of Southeast Asia.

Glossary

arachnid (uh RAK nid) — a group of eight-legged animals, including spiders, scorpions and tarantulas

canopy (KAN uh pee) — the "roof" of upper branches and leaves in a forest

fungi (fung GUY) — a kingdom of living things that includes mushrooms

loris (LOR ihss) — a monkeylike animal of the treetops in parts of Africa and Asia

nocturnal (nok TUR nal) — active at night

prey (PRAY) — an animal that is hunted for food by another animal

species (SPEE sheez) — a certain kind of plant or animal within a closely related group; for example, an *eyelash* viper

INDEX